Anna Marus

Predicting the cost of cab rides using machine learning

Anna Marus

Predicting the cost of cab rides using machine learning

The object of the study is a dataset containing cab rides in New York City for the year 2023

ScienciaScripts

Imprint

Cover image: www.ingimage.com

This book is a translation from the original published under ISBN 978-620-8-17096-7.

Publisher:
Sciencia Scripts
is a trademark of
Dodo Books Indian Ocean Ltd. and OmniScriptum S.R.L publishing group

120 High Road, East Finchley, London, N2 9ED, United Kingdom
Str. Armeneasca 28/1, office 1, Chisinau MD-2012, Republic of Moldova, Europe
Managing Directors: Ieva Konstantinova, Victoria Ursu
info@omniscriptum.com

Printed at: see last page
ISBN: 978-620-8-51152-4

Contents

REFERENCE.

MACHINE LEARNING, TAXI FARE PREDICTION, DATA PREPROCESSING, LEARNING WITH TEACHER, LINEAR REGRESSION, DECISION TREE, RANDOM FOREST, GRADIENT BOUSTING, K-NEAREST NEIGHBOURS METHOD, SUPPORT VECTOR METHOD

The object of the study is a labelled dataset containing taxi rides in New York City for the year 2023.

The aim of the thesis is to perform an intelligent analysis of statistical data of taxi rides in New York City and predict the cost of taxi rides by machine learning.

Machine learning techniques were used to achieve the objective.

The following results were obtained in the thesis:

1) Machine learning methods and tasks are described.

2) Projections of the cost of taxi journeys by different methods are constructed.

The thesis is a practical work. Its results can be used to predict the cost of taxi journeys according to given parameters.

The thesis is complete, the set tasks have been solved in full, there is a possibility of further development of the research.

The diploma work was carried out by the author independently.

INTRODUCTION

The thesis is in the field of machine learning and focuses on training, analysing and predicting data by machine learning using the example of taxi rides in New York City for the year 2023.

The relevance of the topic is that taxis are an integral part of modern society. And forecasting the cost of journeys is of great importance for passengers, taxi drivers and taxi companies. It allows passengers to plan their expenses and avoid unexpectedly high prices. For taxi drivers, it helps them optimise their work, choose more profitable routes and increase their earnings. It helps taxi companies to manage prices, attract more customers and increase their profits. In addition, predicting the cost of taxi rides can be useful for city authorities when planning transport infrastructure and developing urban public transport programmes. Applying machine learning to taxi fare forecasting can take into account many factors such as distance, time of day, weather conditions, demand level and many others to create more accurate and reliable forecasts. Thus, predicting the cost of taxi journeys using machine learning methods is an urgent task that can benefit both individual users and society as a whole.

The purpose of the thesis is to conduct an intellectual
analysing statistics on taxi rides in New York City and predicting
the cost of taxi journeys by machine learning.

In order to achieve the stated goal, the following objectives are set:

- to read statistical data, pre-process them and find new ones
recognitions;
- Analyse the impact of various factors on the cost of a taxi fare and prepare data for machine learning to predict the cost of a journey;
- Using machine learning to explore and create models that predict the cost of taxi journeys.

To write the theoretical part of the thesis we used the methods of analysis, synthesis and classification (description and analysis of machine learning methods, analysis of the cost of a taxi by different indicators). To write the practical part of the thesis were used the method of calculations and measurements, the method of comparison and the method of modelling. The thesis is of a research nature and is a continuation of the course work [7].

CHAPTER 1

THEORETICAL INFORMATION

Nowadays, taxis remain a relevant and in-demand mode of transport. Despite the development of other modes of transport, such as metro, buses, trams and so on, taxis still remain a convenient and efficient way of travelling both in and out of the city.

One of the main reasons for the relevance of taxis is their availability and convenience. Passengers can call a taxi at any time, regardless of weather conditions or time of day. Taxis also offer a personalised service, allowing the passenger to choose the route and time of the journey.

In addition, the development of modern technology and mobile apps has made calling a taxi more convenient and easy. Now you can call a car with just a few taps on your smartphone, it really saves time and simplifies the process.

Taxis also help tourists and travellers who are not familiar with the city to get to their desired destinations quickly and comfortably.

Thus, taxis remain an important mode of transport for modern society due to their convenience, affordability and personalised service.

1.1 The problem of predicting the cost of taxi journeys

The problem with predicting the cost of taxi journeys is that it is a complex regression analysis problem in which many variables and factors that affect the cost of journeys need to be taken into account. Some of the major challenges in using machine learning to predict the cost of taxi journeys include the following:

- Lack of data: accurately estimating the cost of taxi journeys requires access to a large amount of data, access to which may be limited;
- high price variability: the cost of a taxi journey can vary greatly due to time of day, day of the week, season, weather and many other factors. This makes forecasting more challenging;
- accounting for multiple variables: accurate taxi fare estimation requires accounting for multiple variables such as distance, journey time, traffic, weather conditions and demand. This requires sophisticated machine learning models and large amounts of computation;
- accounting for dynamic changes: taxi fares can change in real time depending on supply and demand.

Overall, the problem of predicting the cost of taxi journeys using machine learning techniques is challenging and requires the development of efficient models that can account for multiple variables and dynamically changing conditions.

1.2 Machine learning methods

Machine learning is one of the most relevant and promising areas of data analysis. It is used in various fields, such as search engines, finance, medicine, industrial companies, marketing and transport.

Machine learning helps large corporations and organisations to automate their production processes, improve the quality of decisions made and implemented, increase business efficiency and expand their user base.

Machine learning can also analyse large amounts of data and find hidden patterns that help anticipate future events, optimise production processes and develop useful products and services.

These days, the use of machine learning is becoming more in demand and relevant for any company day by day as it helps in utilising the company's resources efficiently and creating competition in the market.

Machine Learning (ML) is a section of artificial intelligence theory, the subject of which is the search for methods of problem solving by learning while solving similar problems[9, p. 4].

There are many machine learning methods available. The main ones are:

- Supervised Learning (Supervised Learning);
- Unsupervised Learning (Unsupervised Learning);
- Semi-supervised learning (Semi-supervised Learning);
- Reinforcement Learning;
- Transfer Learning (Transfer Learning).

These are the basic types of machine learning and are applied depending on the specific problem and the available data. Each type contains many variations to find solutions to problems of different complexity.

1.2.1 Machine learning with a teacher

Supervised Machine Learning is a method in which the training set of examples contains not only input data but also associated output data. [6, c. 129]

Each training of the model with the teacher consists of the following steps:

1. Data preparation: the data are cleaned, transformed different attributes, and then divided into two samples (training and test);
2. Model selection: suitable models for training are studied and the best one is selected (the main and most common models for training with a teacher are linear and logistic regression, decision trees, random forest, gradient bousting and others);
3. Model training: parameters are tuned to increase accuracy and minimise prediction errors.
4. Model evaluation: the trained model is evaluated with test data, checking its accuracy and its ability to work with other data.

5. model application: the finished model is applied to predict other data.

Learning with a teacher allows the creation of highly accurate models to solve a huge number of problems of varying complexity and is used in many applications.

1.2.2 Machine learning without a teacher

Unsupervised Machine Learning is an algorithm that receives only "signs" as input without any prompting from the teacher[6, p. 134].

The system finds hidden patterns and structures, providing the ability to automatically identify patterns, group data and make predictions.

The best known tasks of teaching without a teacher are:

- clustering (grouping of objects by similar features, highlighting hidden structures in the data);
- dimensionality reduction (reducing the number of attributes while preserving the main characteristics of the data);
- associative rules (finding dependencies between elements of the data set);
- Anomaly detection (identifying unusual and unpredictable values in the data).

Machine learning without a teacher is used to make business decisions, optimise processes and create the latest products and services.

1.2.3 Machine learning with partial teacher involvement

Semi-supervised Machine Learning is a model training method that uses both labelled (correct answer known) and unlabelled (correct answer not given) data. Unlabelled data contains valuable information to help better understand the structure of the data and make predictions more accurate. Various methods are used for this purpose, including clustering tasks, semi-learning and knowledge transfer methods.

Machine learning with partial teacher involvement is used for large amounts of data when data separation is difficult or impossible.

The advantages of the method are obtaining better data, improved models, reduced time and cost of data partitioning. However, there are also limitations, such as the need to use sophisticated algorithms and problems with the interpretability of the resulting models.

1.2.4 Machine learning with reinforcement

Reinforcement Machine Learning is a method in which agents (robots or programmes) are trained to make decisions to increase some reward. The agent learns its environment, makes decisions, these decisions are evaluated and the agents receive a reward for their decision (in the form of a reward or a penalty).

The principle of this method is to train the agent to choose the right behaviour in different situations in order to maximise its final reward. Different methods are

used for this purpose (Q-learning, deep reinforcement learning and actor-critic strategies).
Applications of reinforcement machine learning include board and computer games, robot control, automated trading in financial markets, traffic management, and other tasks where the agent needs to perform a series of actions to achieve goals.
Machine learning with reinforcement learning is one of the most promising areas in artificial intelligence, which is rapidly developing and being used in various fields.

1.2.5 Transfer Machine Learning

Transfer Machine Learning (TL) is a machine learning method in which a model that has been previously trained to perform one task is "fine-tuned" to solve another target task related to the previous one. Training a machine learning model from scratch is typically a labour-intensive and time-consuming process, requiring a large training data set, computational resources of sufficient power, and going through a number of iterations before the model is trained and ready for use. The transfer learning method uses open (the model is freely available) pre trained models, preparing them to solve related problems using new data. [11]
Transfer Learning is used in many industries to automate, optimise and accelerate predictive and decision-making processes.

1.3 Python libraries for machine learning

Python is currently one of the most popular programming languages. Modern Python offers many open source packages and libraries, which makes it convenient for solving problems in machine learning and artificial intelligence.
The best Python libraries for machine learning:

- Tensor Flow is an end-to-end Python machine learning library for performing high quality numerical computations [2];
- Keras is one of the main open source Python libraries written for building neural networks and machine learning projects [2];
- Theano, a machine learning and artificial intelligence library, can handle very large neural networks [2];
- Scikit-learn is a well-known Python machine learning library with a wide range of clustering, regression and classification algorithms [2];
- PyTorch is a fully-ready Python machine learning library with excellent examples, applications and use cases, supported by a strong community [2];
- NumPy is a linear algebra library developed in Python [2];
- Python Pandas is an open source library that offers a wide range of tools for data processing and analysis [2];

- Seaborn is a visualisation library based on Matplotlib [2].

The above-mentioned packages provide completely different functions and tools for data processing, model building and evaluation. The choice of a package to work with depends on the task and level of experience in machine learning.

1.4 Tasks of machine learning

Depending on the type of input parameters and the sought solution, there are several types of problems in machine learning:

- classification tasks: determining the category to which the object belongs;
- Regression problems: predicting numerical values from input data;
- clustering problems: grouping of objects by some similarity;
- Anomaly detection tasks: identifying unusual or anomalous patterns in the data;
- dimensionality reduction tasks: reducing the number of attributes in the original data while preserving all the necessary information;
- tasks Recommendations: Provision of personalised recommendations based on different preferences and behaviours;
- reinforcement tasks: teaching agents to make decisions in their environment and to improve their behaviour based on their evaluations;
- generation tasks: data generation based on training data.

Logistic regression, SVM, Decision Trees, Random Forest, Gradient Boosting and K-Nearest Neighbours are the most commonly used methods in classification tasks. Metrics of accuracy and precision, recall and F1 score are used to evaluate the performance of the models.

Linear regression, random forest regression, decision tree, k-nearest neighbours method, gradient bousting, support vector method and neural networks are most commonly used in regression problems. The metrics of mean absolute error (MAE), mean square error (MSE) and coefficient of determination (R^A2) are used to evaluate the performance of the models.

The k-means method, DBSCAN (Density-Based Spatial Clustering of Applications with Noise), hierarchical clustering, spectral clustering methods and multidimensional scaling algorithms are mainly used in data clustering tasks. Silhouette coefficient and Dunn index metrics are used to evaluate the performance of the models.

Principal component analysis (PCA), t-CNE method, autoencoder method, linear discriminant analysis (LDA) method, random projection method, multidimensional scaling (MDS) method and component analysis (ICA) method are often used in data dimensionality reduction problems.

The recommendation tasks mainly use collaborations filtering, content filtering, hybrid methods, deep learning methods and matrix decomposition methods.

In reinforcement tasks, methods with reinforcement, Q-learning, Deep Q-Networks, Policy Gradient and Actor-Critic are mainly used.
Generative models using neural networks such as Generative Adversarial Networks and Variational Autoencoder, methods based on Recurrent Neural Networks (RNN) and Convolutional Neural Networks (CNN), Autoencoders, deep learning methods such as Long Short-Term Memory and Transformer and reinforcement learning methods are often used in generation tasks.
Since the cost of travelling is a numerical value (i.e. regression), regression algorithms are considered in this paper.

1.4.1 Linear regression

Linear regression is a method that is used to estimate the relationship between the dependent (is a linear combination of the independent variables) and independent variables.
A linear regression model can be represented as an equation:

$$Y = \beta_0 + \beta_1 X_1 + \beta_2 X_2 + \cdots + \beta_n X_n + \varepsilon \qquad (1.1)$$

where Y is the dependent variable, X_1, X_2 ... X_n - independent variables, $\beta_0, \beta_1, \beta_2, \ldots \beta_n$ - regression coefficients, ε - error.
Linear regression (1.1) finds regression coefficients , $\beta_0, \beta_1, \beta_2, \beta_n$, which show the rate of change of the dependent variable Y from the independent variables X_k $(k = 1, \ldots t)$.
The advantages of the method are its simplicity and comprehensibility, interpretability and model efficiency. However, it has disadvantages in the form of linearity, sensitivity to outliers, multicollinearity and limitations on the number of attributes.

1.4.2 Ridge regression

Ridge regression is a regularisation technique that is used in linear regression to combat the problem of multicollinearity (when independent variables in the model are highly correlated with each other) [5]. The basic idea is to add a penalty to the model coefficients to prevent overfitting.
In ridge regression, a penalty term is added to the loss functional, which is the sum of squares of the model coefficients multiplied by the parameter X. Thus, not only the prediction error is minimised, but also the magnitude of the coefficients. The parameter X is selected based on the training data set and allows finding balance between prediction accuracy and model complexity.
Ridge regression helps to improve the generalisability of the model by reducing its tendency to overfitting.

1.4.3 Polynomial regression

Polynomial regression is a technique that is used to estimate the relationship

between a dependent (being a non-linear combination of independent variables) and one or more independent variables. Unlike linear regression, where a linear relationship between variables is assumed, polynomial regression assumes that the relationship can be described by a higher-order polynomial function.
A polynomial regression model can be represented as an equation:

$$Y = \beta_0 + \beta_1 X + \beta_2 X^2 + \cdots + \beta_n X^n + \varepsilon \qquad (1.2)$$

where Y - dependent variable, X - independent variable, $\beta_0, \beta_1, \beta_2, \dots \beta_n$ – regression coefficients, ε - error.
This equation (1.2) is an example of simple polynomial regression. By including additional degrees of X in the equation, more complex nonlinear relationships can be modelled.
An important step in using the polynomial regression method is to choose the optimal degree of the polynomial.

1.4.4 Stochastic gradient descent

Stochastic gradient descent is a machine learning method that optimises model parameters while minimising the loss function. This method is one of the variants of gradient descent and one of the most popular and effective optimisation algorithms in machine learning [8, p. 224].
The essence of the method is to find local maxima or minima of a function using a gradient.
The main difference between stochastic gradient descent and ordinary gradient descent is that stochastic gradient descent computes the gradient on the entire dataset, while stochastic gradient descent computes the gradient on only one random example from the training dataset. This allows for more efficient updating of the model parameters and speeds up the learning process. Therefore, stochastic gradient descent is a more useful method when the dataset is large enough.

1.4.5 Decision tree

A decision tree is a machine learning method for regression and classification tasks that has a tree structure. Any tree necessarily has nodes and edges. The nodes represent conditions on one of the features, and the edges represent the possible outcome of that condition. The tree is constructed by partitioning the data into subgroups.
The advantages of the method are the ease of interpretation of the results, the ability to handle categorical and numerical data, and the automatic handling of missing values. However, there are disadvantages, such as the tendency to overtraining (especially on big data) and instability to changes in the data.

1.4.6 Random forest

Random forest is a machine learning method that uses an ensemble of decision trees for prediction. Each tree is constructed based on a random sample. The predictions of each tree are then averaged and the final prediction is obtained.

The advantages of the method are its ability to handle large amounts of data, its robustness to overtraining, its ability to handle different types of data (categorical and numerical) and its ability to predict non-linear dependencies. However, the method has disadvantages, namely it can be difficult to interpret due to the large number of trees and can take longer to train and predict compared to simpler models.

The random forest method is often used to solve regression problems, especially when the data contain complex non-linear relationships.

1.4.7 Lasso Regression

Lasso regression is a regularisation technique in linear regression that helps to reduce overfitting and improve the generalisability of the model. Lasso regression adds a penalty to the sum of the absolute values of the regression coefficients, while helping to reduce the influence of insignificant features.

The basic idea of Lasso regression is to minimise the sum of the squares of the model prediction errors and the sum of the absolute values of the coefficients with the added regularisation factor X. Selecting the right value of X allows finding a balance between minimising the error and reducing the number of unnecessary parameters.

The use of Lasso regression allows insignificant features to be excluded from the model because some of the feature coefficients are zeroed out. This makes Lasso regression a useful tool for selecting features and improving the interpretability of the model.

The Lasso regression model can be represented as:

$$L = minimize(\sum_{i=1}^{n}(y_i - \hat{y}_i)^2 + \lambda\sum_{j=1}^{p}|w_j|) \quad (1.3)$$

where L is the Lasso model, n number of observations, *p is the* number of features, y_i is the true value of the dependent variable for i-ro observation, $\hat{y}_i$ is the predicted value of the dependent variable for i-ro observation, w_j is the coefficient at j-th feature, λ - is the regularisation factor.

Lasso regression (1.3) is widely used to deal with data that has a large number of features and when it is necessary to select the most important ones.

1.4.8 Gradient bousting

Gradient bousting is a machine learning technique that uses an ensemble of weak models for prediction, which correct the errors of previous models.

Each subsequent model is trained on the errors of the previous models and improves the predictions. For this purpose, gradient descent is used, which minimises the model loss function [10, p. 343].

Gradient bousting, due to its high prediction accuracy, is one of the most popular machine learning methods.

1.4.9 The k-nearest neighbours method

The k-nearest neighbours method is a machine learning algorithm based on the principle of object proximity. It assigns a new object to the class to which most of its k nearest neighbours belong. The method is used for classification and regression.

Principle of the k-nearest neighbours method:

1. k neighbours are chosen;
2. distance from the newly selected object to all objects of the training sample is calculated;
3. k objects with minimum distance are selected;
4. the new object belongs to the class to which most of the k nearest neighbours belong.

The k-nearest neighbours method is easy to implement, but has poor performance on large samples due to the need to calculate distances to all objects in the training sample. To avoid overtraining or undertraining of the model, it is necessary to choose the correct value of the parameter k.

1.4.10 Reference vector method

The support vector method is a machine learning algorithm that searches for an optimal hyperplane that divides the data into two classes. The hyperplane is chosen so that the distance from it to the nearest points of each class (support vectors) is maximised. This method is used for classification and regression tasks.

The advantages of the method are good generalisability, efficiency in dealing with big data and the ability to control the complexity of the model using the regularisation parameter. However, it also has disadvantages, such as the complexity of parameter tuning and the demanding computational resources.

1.4.11 Multilayer perseptron model

A multilayer perseptron is a type of artificial neural network consisting of several layers of neurons, including an input layer, hidden layers, and an output layer. Each neuron in a layer is connected to each neuron in the next layer using weights that determine the strength of the connection between the neurons.

This model can be used to solve classification, regression or function approximation problems. Training is performed by adjusting the weights of neurons using error back propagation algorithms.

Multilayer perseptron is one of the most widespread types of neural networks. It is an effective tool for processing complex data and solving a variety of machine learning problems.

1.5 Conclusion to Chapter 1

Each of the above machine learning methods has its own characteristics and is used for different purposes, depending on the task at hand and the data available. In the taxi industry, machine learning models can analyse large amounts of data and identify patterns that help predict the cost of a ride with high accuracy.

For predicting the cost of taxi journeys, machine learning method with a teacher is most suitable. Since the trip cost is a regression, linear regression, polynomial regression, stochastic gradient descent, decision tree, random forest, gradient bousting, lasso model, k-nearest neighbours method and multilayer perseptron model can be used for prediction.

CHAPTER 2

DATA PREPROCESSING AND ANALYSIS

2.1 Pre-processing of initial data

Yellow taxi trips for the year 2023 are used to predict the cost of taxi rides. "Yellow taxis are authorised to pick up passengers throughout the city, but actually circulate in Manhattan.

2.1.1 Description of initial data

The data used in this paper was collected and provided to the Taxi and Limousine Commission of New York (TLC) by technology providers commissioned under the Taxi Passenger Transportation and Livery Enhancement Programmes (TPEP / LPEP)[3].

The table including all annual yellow taxi trips in New York City for 2023 consists of 18 fields (criteria) and includes over 24,000,000 trips (see Table 2.1).

Table 2.1 Description of initial data

Field Name	Description
VendorlD	Code, indicating the TPEP provider, who provided the recording. 1= Creative Mobile Technologies, LLC; 2= VeriFone Inc.
tpep pickup datetime	The date and time when the meter was switched on.
tpep_dropoff_datetime	The date and time when the meter was switched off.
Passenger_count	The number of passengers in the vehicle. The value is entered by the driver.
Field Name	Description
Trip_distance	The distance travelled in miles shown by the taximeter.
PULocationID	TLC Taxi zone in which the taximeter was switched on
DOLocationlD	TLC Taxi zone where the taximeter has been disabled
RateCodelD	The fare code of the valid trip. 1= Standard rate 2= journeys between New York City boroughs and John F. Kennedy Airport 3= journeys between New York City boroughs and Newark Airport 4= Nassau and Westchester counties 5=Contractual tariff 6=Group travel

Store_and_fwd_flag	Indicates whether the trip record was stored in the vehicle's memory before being sent to the supplier. Y= saved and forwarded N= not saved or forwarded
Payment_type	A code indicating how the passenger paid for the journey. 1= Credit card 2= Cash 3= Free 4= Contract value 5= Unknown 6= Cancelled trip
Fare_amount	Time and distance-based fare calculated using a meter.
Extra	Various additional services and surcharges. These include only rush hour fees of $0.50 and $1 per night.
MTA_tax	Tax of $0.50 per year, which is automatically assessed based on the discount rate used.
Improvement_surcharge	Additional improvement fee of $0.30 or $1.00 per trip.
Field Name	Description
Tip_amount	Tip Amount. This field is automatically filled in for credit card tips and cash tips are not included.
Tolls_amount	The total amount of all fees paid for the trip.
Total_amount	Total amount charged to passengers. Tips included.
Congestion_Surcharge	Total amount charged per trip as a toll in New York City
Airport_fee	$1.25 per transfer. Valid only at La Guardia and John F. Kennedy airports

At the moment the data cannot be analysed and used for machine learning because some values are missing and incorrect data is present. Therefore, it is necessary to process the data.

2.1.2 Reading the raw data

The data used in this paper are 12 parquet files, each containing information on trips for the corresponding month. Each file is a table of 18 columns (criteria) and about 2,000,000 rows (trip records). Python uses the parquet_read function to read parquet data.

You can process the data in a number of ways:

- processing each file separately;
- processing of a combined table with data for the whole year.

Processing each file separately takes a lot of time and is not the best as some data may be lost. Therefore, in this case it is better to collect 12 tables into one and analyse the annual table.

After merging the files, the table contains 24,649,091 rows. Such voluminous tables make it difficult to prepare and process data, so it is necessary to simplify them first and then take them into work.

2.1.3 Optimisation of data types

When loading data into Python using Pandas, types are automatically defined. However, in many cases the output type is not optimised. Moreover, if a numeric column contains missing values, the automatically calculated type will be float. Therefore, it is necessary to optimise the data types.

After examining the dataset, it was found that the table contains the following data types:

- Object (store_and_fwd_flag);
- Int64 (VendorID, passenger_count, RatecodeID, PULocationID, DOLocationlD and payment_type);
- Float64 (trip_distance, fare_amount, extra, mta_tax, tip_amount, tolls_amount, improvement_surcharge, and total_amount, congestion_surcharge and airport_fee);
- Datetime64 (tpep_pickup_datetime and tpep_dropoff_datetime).

In order to reduce memory, data of int64 type were converted to int16 type, and data of float64 type to float32. After the conversion, the tables occupy 3 times less RAM and can be worked with.

2.1.4 Deleting gaps in the table

Often large amounts of data prepared for subsequent work have gaps. In order to be able to use machine learning algorithms that build models on this data, in most cases, these gaps need to be filled.

Missing data can be replaced with specific numeric values or mathematical function values, you can use method fillna() to do this. This method does not modify the structure, it returns a DataFrame created on the basis of an existing one, with NaN values replaced by those passed as an argument.

When checking the table, it turned out that there were missing values in the columns passenger_count, RatecodelD, congestion_surcharge and airport_fee. The missing values were replaced with median values using the midian() function.

2.1.5 Obtaining new traits

One of the most important steps in data preprocessing is the acquisition of new features.

It is known that during peak hours, online taxi aggregators introduce trip multipliers to incentivise available taxi drivers to make trips in areas where demand is strongest. Once demand decreases and there are more available cars in an area, the upward adjustment factor is first reduced and then removed. Thus, to further predict the cost of a taxi, it is necessary to determine the time and date when the journey starts.

The length of the journey is one of the most important attributes in shaping the cost of a taxi ride.

Therefore, new features were obtained from the values of tpep_pickup_datetime and tpep_dropoff_datetime:

- trip_duration - trip duration in minutes (from 0 to 59 minutes);
- pickup_hour - the hour of the trip start (from 0 to 23 hours);
- day_of_week - the day of the week the trip started (1 - Monday, 2 - Tuesday, 3 - Wednesday, 4 - Thursday, 5 - Friday, 6 - Saturday, 0 - Sunday);
- month - month of the trip start (1 - January, 2 - February, 3 - March, 4 - April, 5 - May, 6 - June, 7 - July, 8 - August, 9 - September, 10 - October, 11 - November, 12 - December).

2.1.6 Removal of emissions

An outlier is a data item or object that differs significantly from the rest of the (so-called normal) objects. They can be caused by measurement or execution errors. In other words, outliers are values that are too large or too small in relation to other data.

In taxi trip records, outliers occur quite frequently, as a result of taximeter and equipment malfunction, human error, randomness, or just unique phenomena.

Removing outliers is an important step in preparing data for machine learning. Since outliers and various anomalies represent completely different strongly differing values, their presence can lead to overtraining or a less efficient model.

There are different methods for calculating emissions:

- graphical method (drawing of graphs where the deviation from the norm can be seen);
- Z-score method (search for the degree of deviation of data from the sample mean in standard deviations, outliers are data for which the Z-score value exceeds the threshold value);
- interquartile range method (the difference between the upper and lower quartiles of the data);
- Normal distribution quantile method (if the data are normally distributed, then values outside the range [-3o, 3o] are less than 0.3% of the total number of observations and can be considered outliers).

I chose the interquartile range (IQR) method to clean the data. This method consists of finding the difference between the first (25th percentile) and third (75th percentile) quartile of the data set.
Emissions are values that are outside the range [Q1 - 1.5 * IQR, Q3 + 1.5 * IQR], where IQR is the interquartile range (equal to Q3-Q1).
The interquartile range is commonly used in statistics to determine if there are outliers in the data. If a data value is outside the interquartile range, it is considered an outlier.
This method was applied to the columns extra, trip_distance, trip_duration, tolls_amount and to the target variable total_amount.
For cleaning the data passenger_count, RatecodeID, congestion_surcharge, improvement_surcharge, airport_fee, mta_tax, a graphical method was chosen because the data has a small amount of variation.

2.1.7 Deleting unnecessary columns

To make things easier, we need to remove columns that do not provide useful information for analysing and forecasting the cost of taxi journeys. The following columns can be removed without spoiling the data for the main task:

- VendorlD. The code indicating the supplier who provided the delay record does not carry important information and does not affect the cost of the journey.
- Store_and_fwd_flag. For further work, it is irrelevant whether the trip record was stored in the vehicle memory before sending it to the supplier.
- Payment_type. The payment type does not affect the cost of the journey.
- Fare_amount. The Total_amount field is used to train the neural network, so the amount of the trip according to the meter is not important.
- Tip_amount. Tip is a personal wish of the passenger, so it is not taken into account in the analysis.
- Tpep_pickup_datetime. This column can be deleted after obtaining the necessary data about the duration, time, day and month of the trip.
- Tpep_dropoff_datetime. This column can be deleted after obtaining the required trip duration data.
- Tolls_amount. After removing the outliers, all values in this column became 0. It does not carry any information, so it can be deleted.

Thus, after deleting all unnecessary data, there are 15 columns left in the table.

2.1.8 Coding of categorical attributes

Categorical feature encoding is the process of converting categorical data into numerical values that can be used by machine learning algorithms [1]. Categorical feature encoding is an important step in any data preprocessing, as most machine learning algorithms work only with numbers.
There are several ways of coding categorical attributes:

- One-Hot Encoding (Dummy Encoding): each unique category is converted into a separate binary column that indicates the presence or absence of that category;
- Label Encoding: each unique category is converted to a numeric value;
- Target Encoding: each category value is replaced by the average value of the target variable for that category.

The processed data has one column with categorical attributes, RatecodeID (trip type). To encode this column, I chose the One-Hot Encoding method. Six new attributes were created, where 6 is the number of trip types. Each new feature is a binary characteristic feature of the nth category.

2.1.9 Data normalisation

Numerical data normalisation is the process of bringing numerical values to a certain range or scale to make them comparable and improve the performance of machine learning algorithms.

Normalising numerical data can improve the convergence of machine learning algorithms, reduce the impact of outliers and simplify the interpretation of results.

The most common methods for normalising numerical data:

- min-max normalisation: bringing values into the range from 0 to 1, by subtracting the minimum value and dividing by the difference between the maximum and minimum values;
- z-normalisation (standardisation): transforming values so that the mean is 0 and the standard deviation is 1;
- robust normalisation: transforming values so that they are robust to outliers by subtracting the median and dividing by the interquartile range;
- Logarithmic normalisation: transforming values by taking their logarithm;
- Binarisation: conversion of numeric values into binary values (0 or 1) based on a defined threshold;
- length normalisation: transforming vectors so that their length is equal to 1;
- categorical normalisation: converting categorical data into numeric data by assigning unique numeric values to them;
- Mean normalisation: converting values by subtracting the mean and dividing by the standard deviation.

The standardisation method (StandardScaler) was chosen for the data to be processed.

2.2 Data analysis

2.2.1 Visualisation of cost by length and distance of journeys

A taxi journey is usually based on a taximeter. The cost per kilometre is different for all companies. Thus, the price of a taxi ride depends on two factors:

the length of the route (Figure 2.1) and the cost per kilometre in the selected company. But the type of journey plays an important role, because the same length and distance of journey can have completely different prices due to the fare.

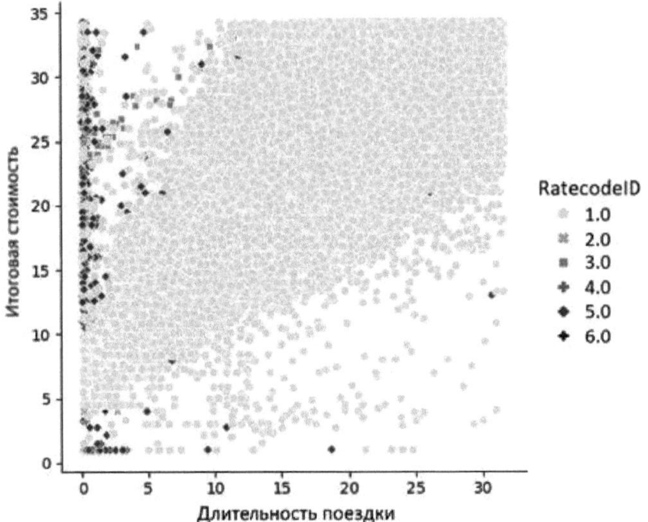

Figure 2.1 Visualisation of trip cost versus duration by attribute RatecodelD (trip type)

It can be seen that the cost increases proportionately with the length of the journey.

2.2.2 Effect of time of day on cost and number of journeys

The number of trips per day varies greatly (Figure 2.2). This creates the so-called rush hour.

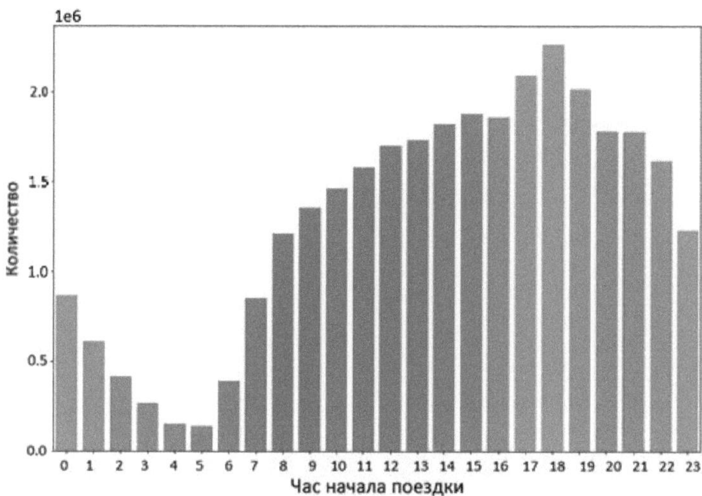

Figure 2.2 Number of journeys by hour

It can be seen that the highest demand for taxis is in the evening (17:00-19:00), when people are travelling home from work. The lowest demand for taxis is late at night (03:00-05:00).

The cost of journeys along the same route also varies throughout the day (Figure 2.3). The difference in the cost of journeys is mainly influenced by demand, i.e. the number of taxi orders.

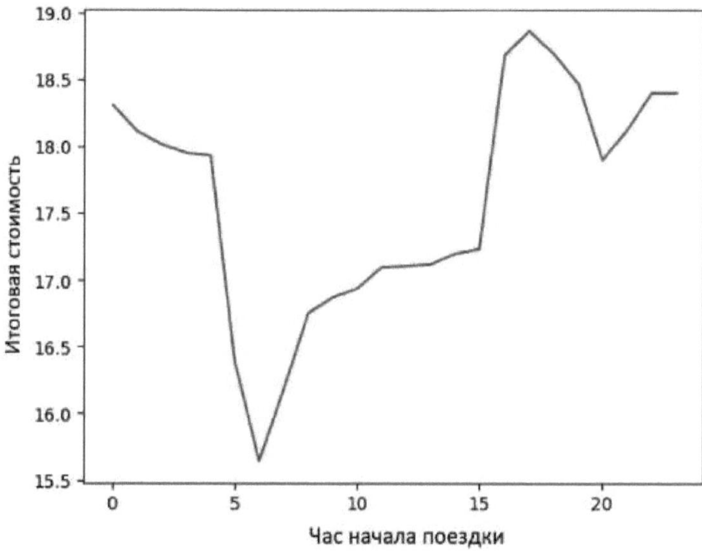

Figure 2.3 Cost of journeys depending on the starting hour of the journey

2.2.3 Median values of cost, distance and trip duration depending on the type of RatecodeID

The depiction of median values of cost, distance and trip duration will help to determine the dependence of these data on trip type (Figure 2.4).

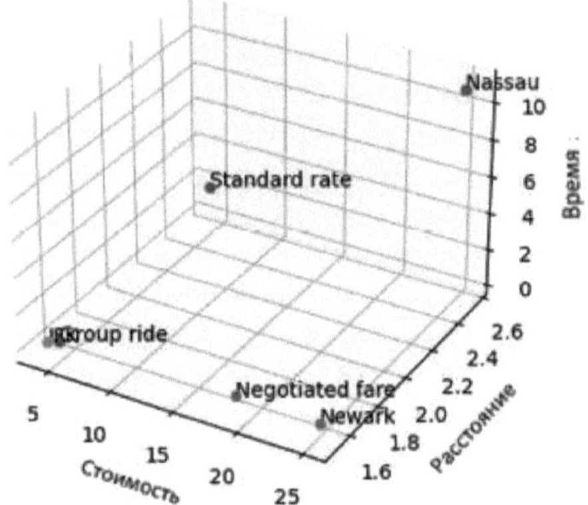

Figure 2.4 Median values of cost, distance and trip duration by RatecodeID type

We can conclude that the most expensive trips are those between New York City boroughs and Newark Airport. The cheapest trips are those between New York City boroughs and John F. Kennedy Airport.

2.3 Result of data preprocessing and analysis

After data preprocessing, the table consists of 18 columns and over 22,000,000 rows (Figure 2.5).

0	1	2	3	4	5	6	7	8	9	10
-0.437995	-0.514804	1.087451	1.074225	1.120034	0.023097	0.0124	0.201683	-0.964998	-2.454054	-0.456143
-0.437995	-0.514804	1.102636	1.059964	1.120034	0.023097	0.0124	0.201683	-0.480356	-2.454054	-0.456143
-0.437995	-1.033178	1.087451	1.059964	1.120034	0.023097	0.0124	0.201683	-0.708786	-2.454054	-0.456143
-0.437995	-0.860387	1.087451	-0.180739	-1.161574	0.023097	0.0124	-4.958279	-0.955737	-2.454054	-0.456143
-0.437995	-1.525633	-2.420271	0.418221	-1.161574	0.023097	0.0124	-4.958279	-1.690418	-2.454054	-0.456143
1.562150	-0.843108	-0.036235	-1.364397	-1.161574	0.023097	0.0124	0.201683	-0.662483	1.244473	0.087921
-0.437995	0.029488	-0.339933	-1.050656	-1.161574	0.023097	0.0124	0.201683	-0.708786	1.083667	0.087921
-0.437995	0.331873	1.057081	-0.038129	-1.161574	0.023097	0.0124	0.201683	-0.332185	1.083667	0.087921
-0.437995	0.625618	1.057081	-1.991879	-1.161574	0.023097	0.0124	0.201683	-0.347620	1.244473	0.087921
-0.437995	-1.041818	1.087451	1.074225	-1.161574	0.023097	0.0124	0.201683	-1.353947	1.244473	0.087921

Figure 2.5 View of the table after preprocessing

Random partitioning, cross-validation or temporal partitioning methods are used to divide the data into training and test samples.

The most common method is random partitioning, in which the data are divided into two samples randomly. Typically, a 70/30 or 80/20 ratio is used for the training and test samples respectively.

After dividing the data into training and test samples, we can proceed to train the model on the training sample and evaluate its quality on the test sample.

2.4 Conclusions to Chapter 2

One of the most important steps in machine learning is data preprocessing.

Data preprocessing included optimising data types, removing table omissions, obtaining new features, removing outliers, removing redundant columns, coding categorical features, normalising the data and dividing the dataset into training and test samples.

All these steps help to improve data quality, avoid distortions and incorrect results, identify dependencies and select the most relevant features for modelling.

CHAPTER 3

MACHINE LEARNING

3.1 Hyperparameters and model parameters

Machine learning models include two types of parameters: *hyperparameters* (external to the model, values cannot be estimated from data) and *parameters* (internal to the model, values are estimated from data).

Hyperparameters are the parameters of the model that are tuned before the training process begins. They determine the very structure of the model and the way it is trained, and are crucial for achieving its high performance and accuracy. Hyperparameters should be tuned so that the model can optimally solve the training task.

There are several approaches to optimise hyperparameters: GridSearch, RandomizedSearch, Bayesian optimisation, etc.

The GridSearch approach consists in creating a grid of hyperparameters (several values are fixed for each hyperparameter) and training/testing the model on each of their possible combinations. This approach performs a complete search of all specified combinations of hyperparameters, but it can be very resource-intensive if the number of hyperparameters is large.

The RandomizedSearch approach is to create a grid of hyperparameters and train/test the model only on some randomly selected combination of these hyperparameters. This approach is more time efficient, but may miss some combinations, possibly the best ones.

Due to the large amount of raw data, it is not feasible to use the above methods. Therefore, the Bayesian hyperparameter optimisation algorithm BayesSearchCV is applied with taxi trip data.

BayesSearchCV is part of the scikit-optimize library, which provides tools for optimising hyperparameter models.

BayesSearchCV allows you to find optimal values of model hyperparameters by minimising a loss function or maximising a quality metric on a training dataset. It uses Bayesian optimisation algorithm to efficiently find optimal values of hyperparameters.

Advantages of BayesSearchCV include a more efficient search for optimal hyperparameter values compared to classical methods such as grid search or random search. It also automatically adapts to the results of previous iterations, allowing faster convergence to the optimal solution.

BayesSearchCV is a powerful tool for optimising parameters of machine learning models and is used for different tasks: classification, regression clustering and others. It helps to improve the quality of models, speed up the

development and training process.
Model parameters (internal to the model) - parameters that are changed and optimised in the process of model training, their final values are the result of model training.
Next, let's look at the machine learning methods implemented in the thesis.

3.2 Linear regression

LinearRegression is a linear regression method from the scikit-learn (sklearn) library in Python. This method is used to model the relationship between one or more independent variables and the dependent variable by fitting a linear function to the data.
In LinearRegression we can set the hyperparameters fit_intercept (intercept score, default True), copy_X (copy X, default True), n_jobs (number of jobs that are used for computation, default None) and positive (whether the coefficients should be positive, default False). When searching for hyperparameters, it turned out that the best combination for this model is the combination that is set by default.
Training the data using the LinearRegression machine learning model (Figure 3.1) resulted in a mean square error (MSE) of 0.483 and a mean absolute error (MAE) of 0.415.

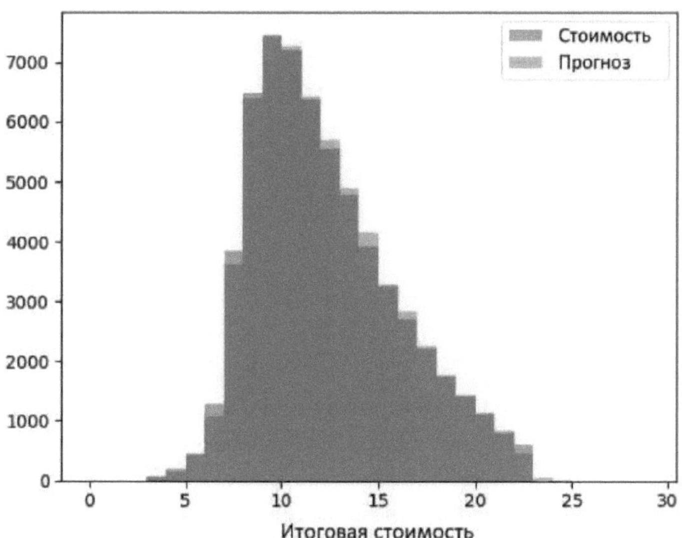

Figure 3.1 Histogram of prediction using linear regression model

3.3 Ridge regression

Ridge regression is a linear regression method with l2-regularisation to prevent overfitting of the model. This method is used when multicollinearity is present in the data, i.e. when features are highly correlated with each other.

Various parameters are fed into the Ridge class, the main ones being:

- alpha: parameter controlling the strength of regularisation (the larger the alpha, the stronger the regularisation), default is 1.0 [4];
- fit_intercept: intercept score, defaults to True [4];
- copy_X: copy X, default True [4];
- max_iter: maximum number of iterations, defaults to None [4];
- tol: solution accuracy, default 0.0001 [4];
- solver: type of solver (auto, svd, cholesky, Isqr, sparse_cg, sag, saga, Ibfgs), default is auto [4].

Training the data using the linear least squares method with L2-peryaapn3aunen Ridge with no specified hyperparameters (Figure 3.2) resulted in a mean square error (MSE) of 0.468 and a mean absolute error (MAE) of 0.413.

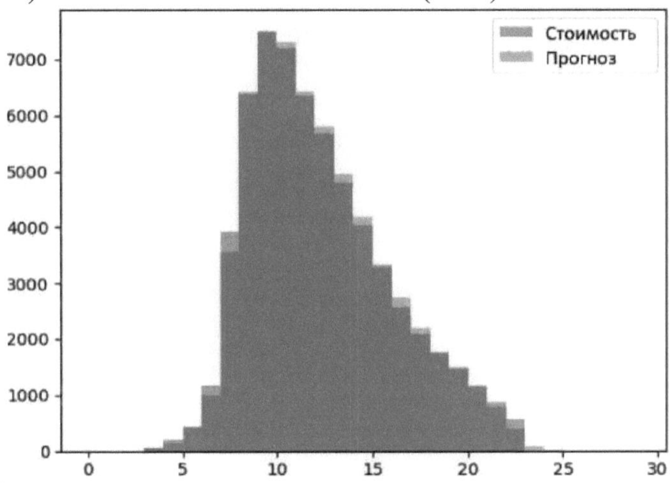

Total cost

Figure 3.2 Prediction histogram using the ridge regression model

3.4 Lasso Regression

Lasso is a linear regularisation method that is often used for feature selection in machine learning models. It adds a penalty to the usual loss function in the form of a sum of absolute values of the model coefficients.

Various parameters can be submitted to the Lasso class, the main ones are:

- alpha: parameter controlling the strength of regularisation (the larger the alpha, the stronger the regularisation), default is 1.0 [4];
- fit_intercept: intercept score, defaults to True [4];
- precompute: use the computed Gram matrix, default is False [4];
- copy_X: copy X, default True [4];
- max_iter: maximum number of iterations, default is 1000 [4];

- tol: solution accuracy, default 0.0001 [4];
- selection: coefficient update (cyclic, random), default cyclic [4].

Training the data using a linear method with l1- Lasso regularisation (Figure 3.3) resulted in a mean square error (MSE) of 1.972 and a mean absolute error (MAE) of 1.045.

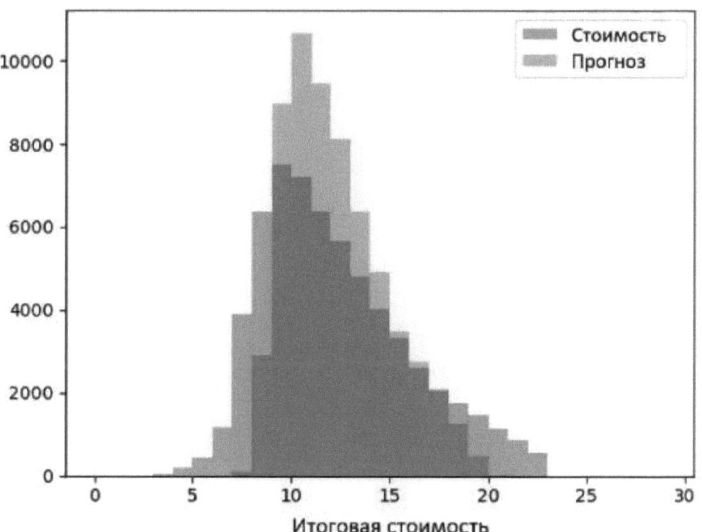

Figure 3.3 Prediction histogram using the Lasso model

3.5 Stochastic gradient descent

SGDRegressor is a regression model based on the stochastic gradient descent method. It works by minimising losses.

SGDRegressor's main parameters:

- loss: the loss function to be minimised (squared_error,huber, epsilon_insensitivity or squared_epsilon_insensitivity), default is squared_error [4];
- penalty: defines the regularisation of the model to prevent overtraining (l2, l1, elasticnet, None), default l2 [4];
- alpha: parameter controlling the strength of regularisation (the larger the alpha, the stronger the regularisation), default 0.0001 [4];
- fit_intercept: intercept score, defaults to True [4];
- max_iter: maximum number of iterations to train the model, default is 1000 [4];
- tol: solution accuracy, default 0.001 [4];
- learning_rate: parameter that defines the learning rate of the algorithm (constant, optimal, invscaling or adaptive), default invscaling [4].

Training the data using the SGDRegressor model with the specified hyperparameters (Figure 3.4) resulted in a mean square error (MSE) of 0.470 and a mean absolute error (MAE) of 0.414.

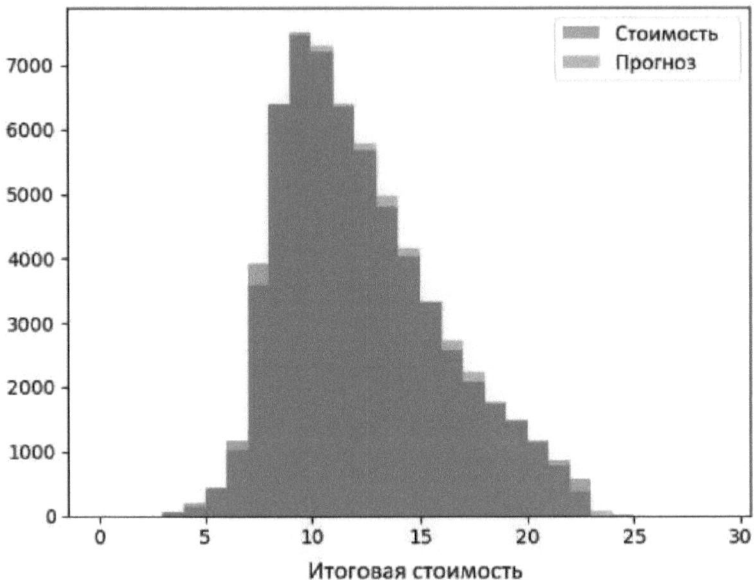

Figure 3.4 Histogram of prediction using the stochastic gradient descent model

3.6 Decision tree

DecisionTreeRegressor is a machine learning model based on a decision tree algorithm that builds a tree by partitioning the data into subgroups and predicting the values of the target variable for new observations.

DecisionTreeRegressor has several parameters that can be adjusted to improve the quality of the model. These are maximum tree depth, minimum number of observations in a leaf, and split criterion. You can also use post-processing techniques such as tree pruning or tree ensembling to improve the predictive power of the model.

Training the data using the DecisionTreeRegressor machine learning model (Figure 3.5) resulted in a mean square error (MSE) of 0.720 and a mean absolute error (MAE) of 0.358.

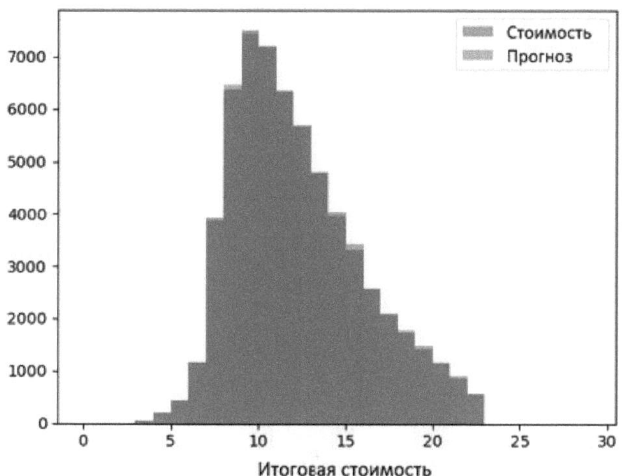

Figure 3.5 Histogram of prediction using the decision tree model

3.7 Random forest

RandomForestRegressor is a machine learning algorithm based on combining a large number of decision trees to obtain a more accurate prediction.

The way RandomForestRegressor works is that it creates an ensemble of decision trees, each of which is trained on a random subsample of data and using random features. Then, for each feature, the model averages the predictions of all trees to produce a final prediction.

Training the data using the RandomForestRegressor machine learning model (Figure 3.6) resulted in a mean square error (MSE) of 0.362 and a mean absolute error (MAE) of 0.287.

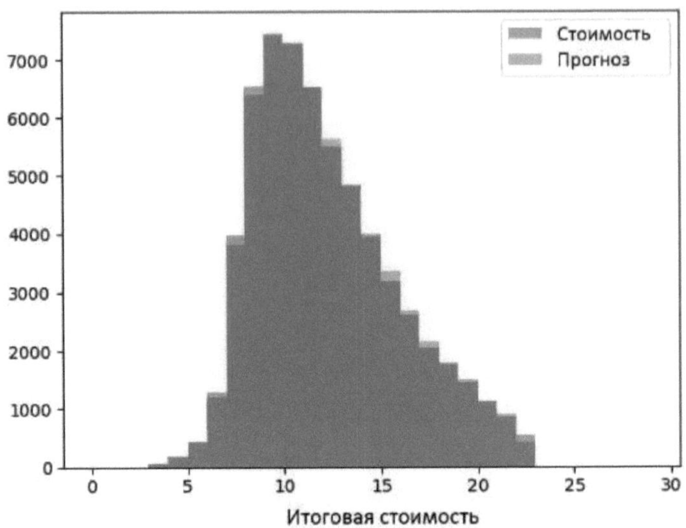

Figure 3.6 Histogram of the prediction using the random forest model

This model showed one of the best results, so we will try to improve it.

The main parameters that can be configured when using RandomForestRegressor include the number of trees in the ensemble (n_estimators), the maximum depth of trees (max_depth), the minimum number of objects in the leaf (min_samples_leaf), and others.

The Bayes algorithm determined that the best hyperparameters are n_estimators=250, max_depth=15, min_samples_split=10, max_features=10, oob_score= True, verbose=250.

Training the data using the RandomForestRegressor model with the specified hyperparameters (Figure 3.7) resulted in a mean square error (MSE) of 0.338 and a mean absolute error (MAE) of 0.270.

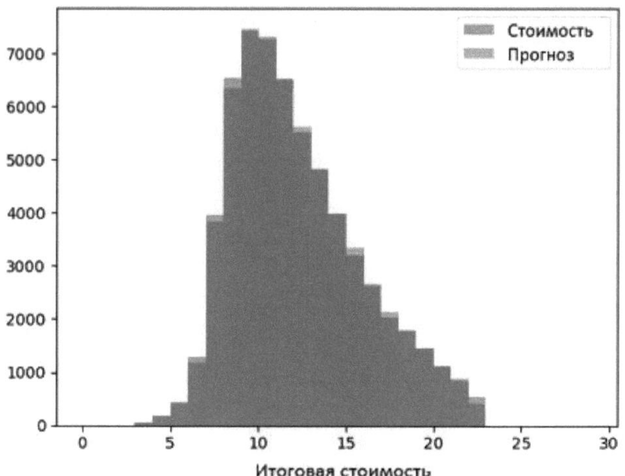

Figure 3.7 Prediction histogram using a random forest model with given hyperparameters

To test the prediction, a graph of the cost per hundred journeys was additionally plotted (Figure 3.8).

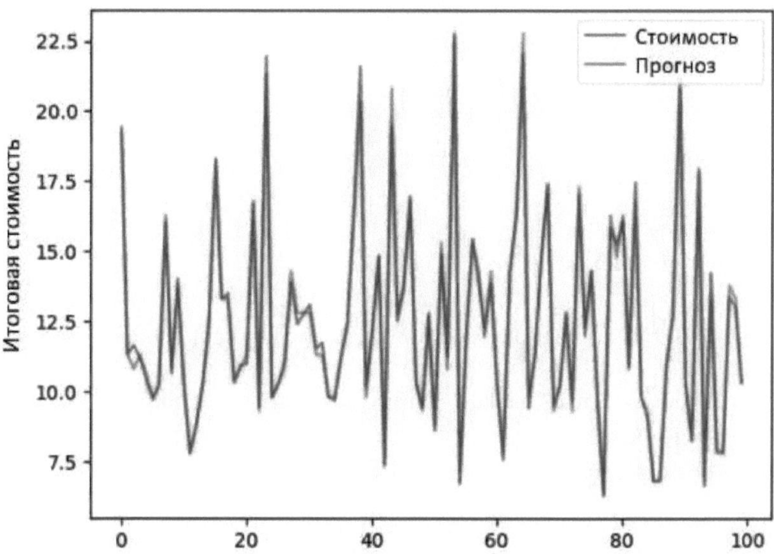

Figure 3.8 Prediction plot of one hundred trips using the random model

scaffolding

3.8 Gradient bousting

GradientBoostingRegressor is a machine learning method based on building a sequence of weak models, each of which corrects the errors of the previous model.

Basic principles of GradientBoostingRegressor operation:

1. initialising the base model;
2. adding the next model to the ensemble, minimising the error of the previous ones (gradient descent is used for this purpose);
3. updating the weights of all models in such a way as to reduce the prediction error;
4. the process of adding new models continues until the stopping criterion is reached.

Training the data with the GradientBoostingRegressor machine learning model (Figure 3.9) resulted in a mean square error (MSE) of 0.360 and a mean absolute error (MAE) of 0.306.

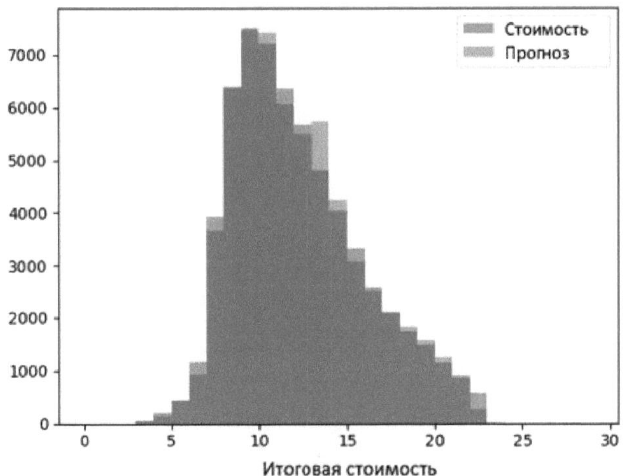

Figure 3.9 Histogram of prediction using the gradient bousting model

This model showed one of the best results, so we will try to improve it.

The main parameters that can be configured when using GradientBoostingRegressor:

Here are some basic parameters of GradientBoostingRegressor:

- loss: loss function, default squared_error [4];
- n_estimators: number of trees to be built during training, default is 100 [4];
- learning_rate: learning rate, default is 0.1 [4];
- max_depth: maximum depth of each decision tree, default is 3 [4];
- min_samples_split: minimum number of samples required to split an internal tree node, default is 2 [4];

• min_samples_leaf: minimum number of samples in a leaf node, default 1, and others [4].

The Bayes algorithm determined that the best hyperparameters are loss='huber', learning_rate=0.7, n_estimators=500, min_samples_split=10, min_samples_leaf= 10, verbose=500.

Training the data using the GradientBoostingRegressor model with the given hyperparameters (Figure 3.10) resulted in a mean square error (MSE) of 0.343 and a mean absolute error (MAE) of 0.261.

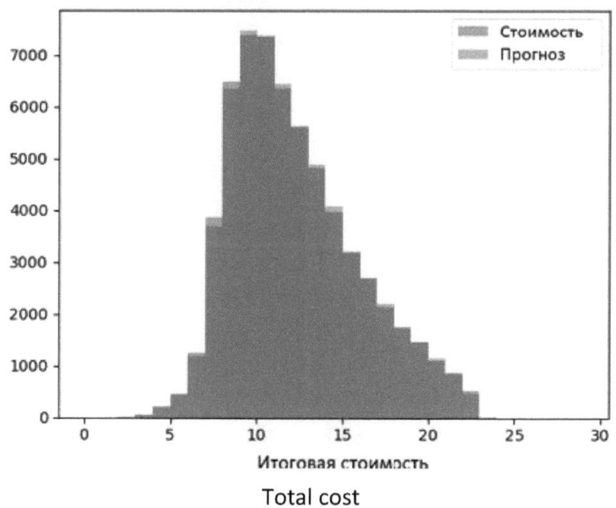

Figure 3.10 Histogram of the prediction using the gradient model bousting model with given hyperparameters

To test the prediction, a graph of the cost per hundred journeys was additionally plotted (Figure 3.11).

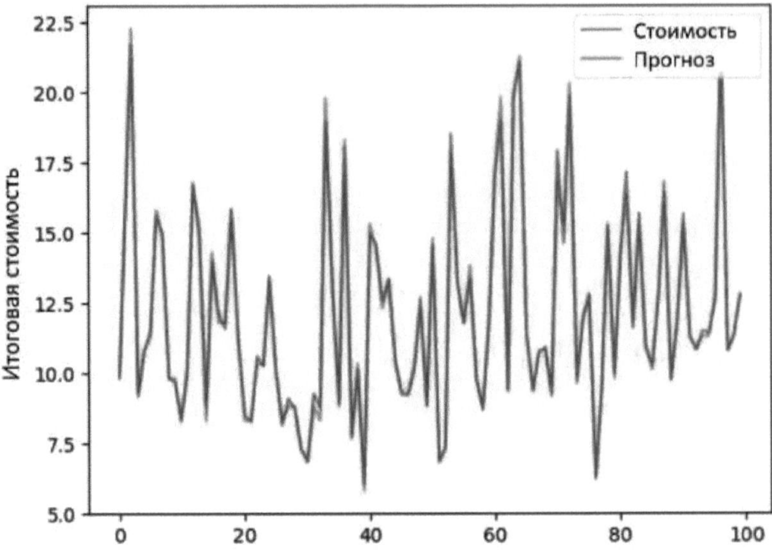

Figure 3.11 Graph of predicting one hundred trips using the gradient bousting model

3.9 The k-nearest neighbours method

KNeighborsRegressor is a regression model that searches for k nearest neighbours for each point in the test dataset and averages their target values to predict the value of the target variable for that point.

The following parameters can be set to use KNeighborsRegressor:

- n_neighbors: number of nearest neighbours to be used for prediction, default is 5 [4];
- weights: weights that can be assigned to each neighbour depending on their distance to the prediction point (uniform or distance), by default uniform [4];
- metric: distance metric used to determine nearest neighbours, defaults to minkowski [4];
- algorithm: the algorithm used to compute nearest neighbours (auto, ball_tree, kd_tree, brute), defaults to auto, and others [4].

Training the data using the KNeighborsRegressor machine learning model (Figure 3.12) resulted in a mean square error (MSE) of 0.434 and a mean absolute error (MAE) of 0.358.

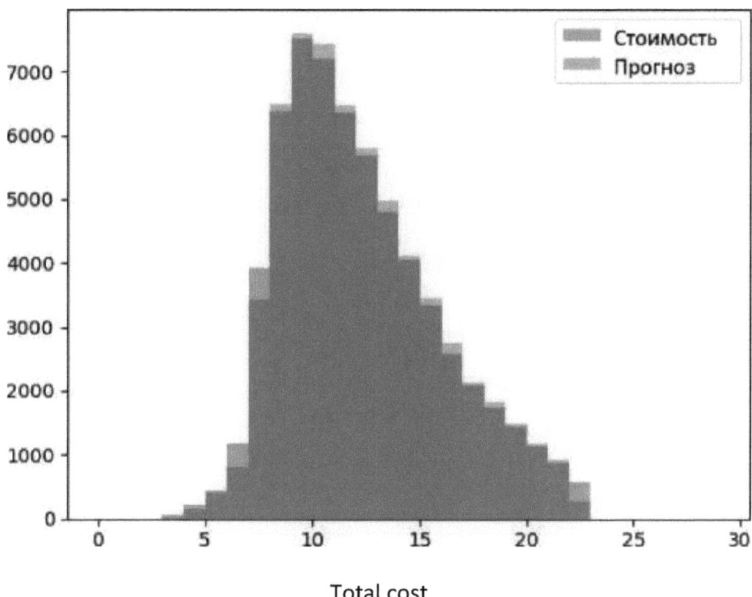

Figure 3.82 Histogram of prediction using the k-nearest neighbour model neighbours

3.10 Reference vector method

SVR is a machine learning method based on optimising a separating hyperplane by considering the width of the gap within which the data points should be located.

Training the data using the SVR machine learning model (Figure 3.13) resulted in a mean square error (MSE) of 0.344 and a mean absolute error (MAE) of 0.265.

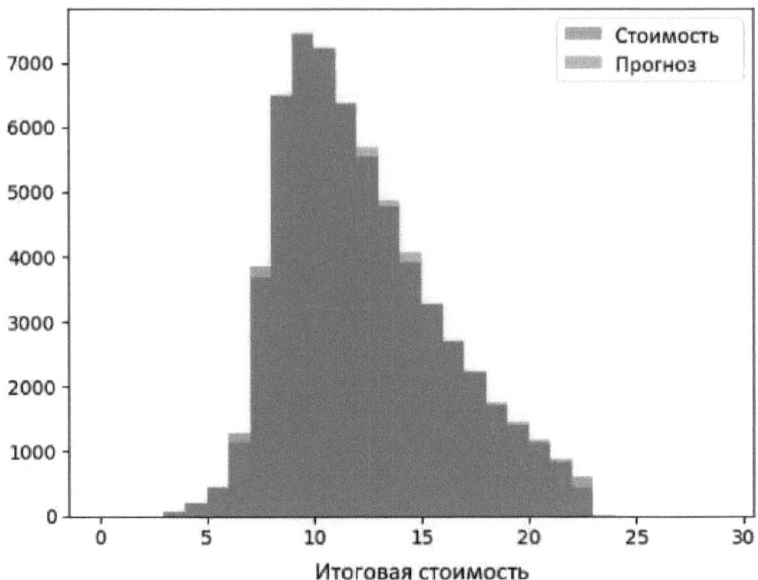

Figure 3.13 Histogram of prediction using support vector model

This model showed one of the best results, so we will try to improve it.

In the SVR method, there are several parameters that can be tuned to achieve better model performance:

- C: regularisation parameter (large values of C mean tighter gaps and less error tolerance on the training data, which can lead to overfitting, while small values allow more points to be inside the epsilon tube), default 1 [4];
- epsilon: epsilon tube width parameter, default 0.1 [4];
- kernel: kernel (linear, poly, rbf, sigmoid or precomputed), default rbf, and others [4].

The Bayes algorithm determined that the best hyperparameters are tol=1e-5, C=16.

Training the data with the GradientBoostingRegressor model with the given hyperparameters (Figure 3.14) resulted in a mean square error (MSE) of 0.340 and a mean absolute error (MAE) of 0.260.

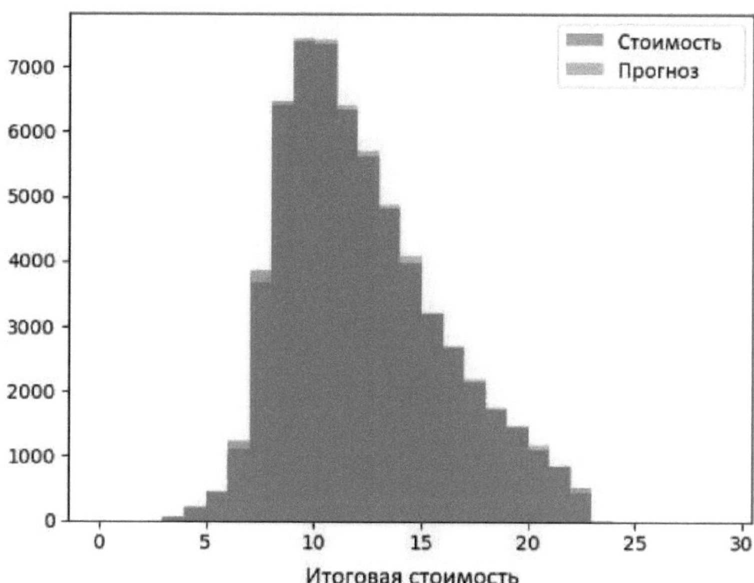

Figure 3.14 Histogram of prediction using support vector model with given hyperparameters

To test the prediction, a graph of the cost per hundred journeys was additionally plotted (Figure 3.15).

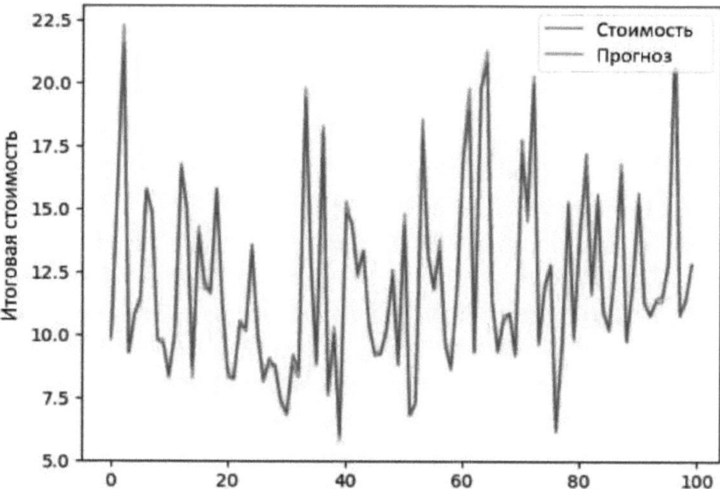

Figure 3.15 Graph of predicting one hundred trips using the support vector model

3.11 A model of a multilayer perseptron

MLPRegressor, a neural network model, is a multilayer perceptron consisting of several layers of neurons, including an input layer, hidden layers and an output layer. Error back propagation method is used to train the model to minimise the prediction error of the model.

Training the data using the MLPRegressor machine learning model (Figure 3.16) resulted in a mean square error (MSE) of 0.341 and a mean absolute error (MAE) of 0.274.

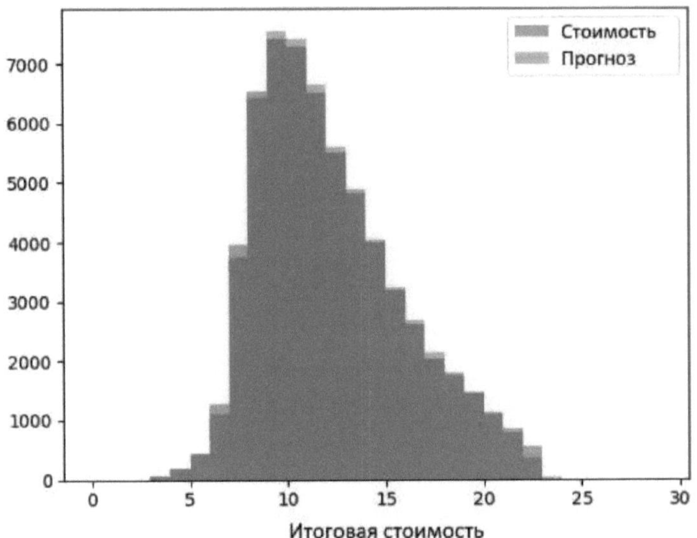

Figure 3.16 Prediction histogram using the multilayer perseptron model

This model showed one of the best results, so we will try to improve it.

The main parameters that can be configured when using the MLPRegressor method:

- hidden_layer_sizes: the number of layers and the number of neurons in each layer of the neural network, default (100,) [4];
- activation: activation function for hidden layers (identity, logistic, tanh or relu), default relu [4];
- solver: algorithm for training the neural network (lbfgs, sgd or adam), default is adam [4];
- alpha: regularisation parameter, default 0.0001 [4];
- learning_rate: neural network learning rate (constant, invscaling or adaptive), by default constant [4];
- max_iter: maximum number of iterations to train the model, by default 200, and others [4].

The Bayes algorithm determined that the best hyperparameters are hidden_layer_sizes=(300,150,50,), activation-logistic', solver='adam', max_iter=500.

Training the data with the MLPRegressor model with the given hyperparameters (Figure 3.17) resulted in a mean square error (MSE) of 0.331 and a mean absolute error (MAE) of 0.269.

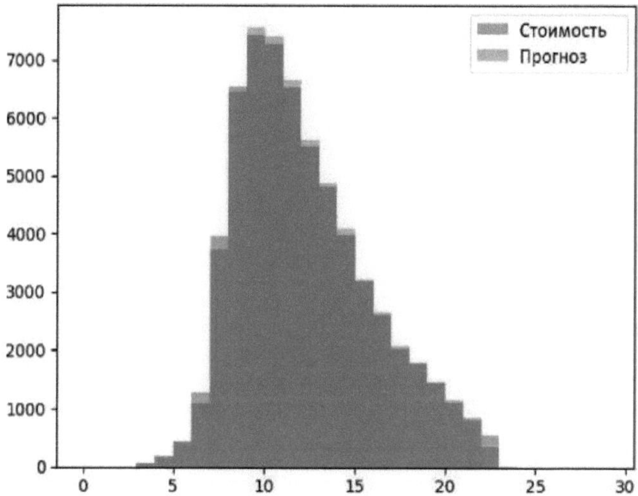

Total cost

Figure 3.17 Prediction histogram using a multilayer perseptron model with given hyperparameters

To test the prediction, a graph of the cost per hundred journeys was additionally plotted (Figure 3.18).

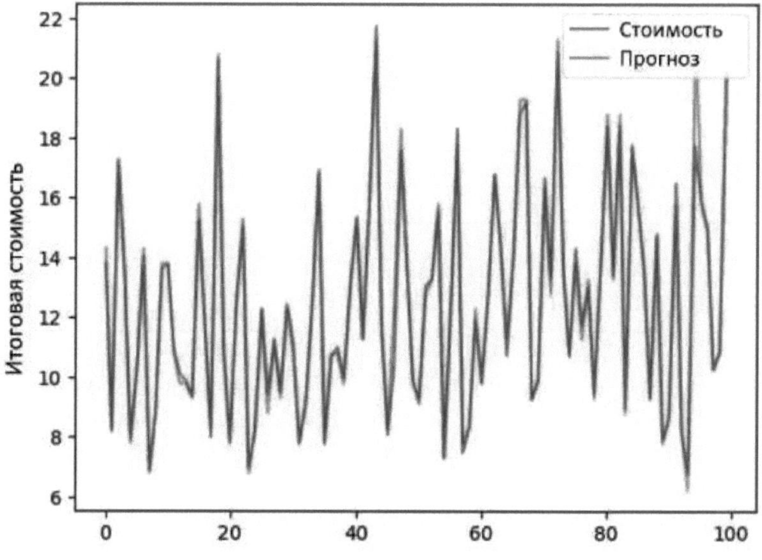

Figure 3.18 Graph of predicting one hundred trips using the multilayer perseptron model

3.12 Comparative analysis of models

The metrics most commonly used to evaluate the prediction performance of regression models are mean absolute error (3.1), mean squared error (3.2) and coefficient of determination (3.3):

$$MAE = \frac{1}{n}\sum_{i=1}^{n} |y_i - y_i^*| \quad (3.1)$$

$$MSE = \sqrt{\frac{1}{n}\sum_{i=1}^{n} (y_i - y_i^*)^2} \quad (3.2)$$

$$R^2 = 1 - \frac{\sum_{i=1}^{n}(y_i - y_i^*)^2}{\sum_{i=1}^{n}(y_i - y_i^{**})^2} \quad (3.3)$$

where y_i *is the* actual value for the i-th observation, y_i^* is the predicted value for the i-ro observation, y_i^{**} is the arithmetic mean of all correct answers, *n is the* number of observations.

To better evaluate the performance of the models, it is necessary to find the above metrics for each model built and compare their results (Table 3.1).

Table 3.1 Comparative analysis of models

Model/Metric	*MAE*	*MSE*	R^2
LinearRegression	0.415	0.483	0.966
Ridge	0.413	0.468	0.966
Lasso	1.045	1.972	0.858
SGDRegressor	0.414	0.470	0.966
DecisionTreeRegressor	0.358	0.720	0.948
RandomForestRegressor	0.270	0.338	**0.976**
GradientBoostingRegressor	0.261	0.343	0.975
SVR	**0.260**	0.340	**0.976**
KNeighborsRegressor	0.358	0.434	0.969
MLPRegressor	0.269	**0.331**	**0.976**

The coefficient of determination measures the amount of variance in the predictions of machine learning models. This measure is also referred to as a measure of model quality. The closer the value of R^2 is to one, the better the model explains the data and, conversely, if the value is close to zero, the predictions are a constant prediction.

All models showed good results. The support vector model (SVR) showed the lowest mean absolute error, the multilayer perseptron model (MLPRegressor) showed the lowest mean square error, and the coefficient of determination was the best for RandomForestRegressor, SVR and MLPRegressor models.

3.13 Conclusion to Chapter 3

Linear and ridge regression, Lasso regression, stochastic gradient descent, decision tree, random forest and gradient bousting methods, support vector methods, k-nearest neighbours and multilayer perseptron model were investigated.

For taxi trip data, the random forest, gradient bousting, support vector method and multilayer perseptron models give the best predictions. This may be due to the fact that the cost of journeys depends on several factors, which when combined in different ways give different cost of a journey. The other models also show good results.

Hyperparameters were selected for four models with higher prediction accuracy (RandomForestRegressor, GradientBoostingRegressor, SVR and MLPRegressor). The SVR model (support vector method) showed the best result, but this model requires high computer performance.

CONCLUSION

In the course of the thesis, various methods for solving the regression problem related to building machine learning models were investigated: linear regression, ridge regression, polynomial regression, Lasso regression, stochastic gradient descent, decision tree, random forest, gradient bousting, support vector method, k-nearest neighbours method and multilayer perseptron model.

The raw data was intelligently analysed and preprocessed for further machine learning.

The result was models trained on the original dataset. The quality of the models was assessed by MSE and MAE loss functions.

The findings revealed that the best machine learning methods for predicting the cost of taxi rides are Random Forest, Gradient Busting, Support Vector Method and Multilayer Perseptron Model. After fitting the hyperparameters, the mean absolute error was 0.26 and RMS error was 0.33.

LIST OF REFERENCES

1. 4BRAIN. Lesson 3: Data processing for machine learning [Electronic resource]. resource] - Mode Access: https://4brain.ru/aibasics/data.php7ysclidHvldcddgx572100754. Date of access - 28.04.2024

2. PYTHONIST. Top 8 Python libraries for machine learning and artificial intelligence [Electronic resource] - Mode of access: https://pythonist. ru/top-8-bibliotek-python-dlya-mashinno go-obucheniya-i- iskusstvennogo-intellekta/?ysclid= lvjfdiinpv730615971. Date of access - 22.04.2024

3. TLC Trip Record Data [Electronic Resource] - Mode of Access: https://www1. nyc.gov/site/tlc/about/tlc-trip-record-data.page . Access Date - 02.04.2024

4. Scikit-learn. Machine learning in Python [Electronic resource] - Mode of access: https://scikit-learn.org/stable/index.html . Date of access - 08.05.2024

5. Big Russian Encyclopaedia. Grebnevaia regression [Electronic resource] - Access mode: https: //bi genc. ru/c/grebnevaia-regressiia-9aa84b. Date of access - 16.04.2024

6. Goodfellow, J., Bengio, I., Courville, A. Deep Learning. 2nd ed., revised. - Moscow: DMK Press, 2018.

7. Marous A. INTELLIGENT ANALYSIS OF STATISTICAL DATA OF TAXI JOURNEYS, 2023

8. Shafrin Y. Information technologies: in 2 parts / Y. Shafrin. - Moscow: Binom, Laboratory of Knowledge, 2004. - Part 1. Fundamentals of informatics and information technologies. - Mode access: http: //intsys.msu.ru/staff/mironov/machine learning vol 1.pdf. Date of access - 24.04.2024

9. Watt, D., Borhani, R., Katsaggelos, A. Machine learning: fundamentals, algorithms and application practice. - S.-Pb.: BHV-Peterburg, 2022.

10. Flach, P. Machine Learning. The science and art of building algorithms that extract knowledge from data. - Moscow: DMK Press, 2015.

11. What is transfer learning? [Electronic resource] - Mode of access: . https://aws.amazon.com/ru/what-is/transfer-learning/Date of access - 28.05.2024

APPENDIX

Programme code

Below is the code of the Python work

```
import pandas as pd

import numpy as np

import glob

import random

import seaborn as sns

import matplotlib.pyplot as plt

import scipy.stats as stats

from sklearn.preprocessing import OneHotEncoder

from sklearn.preprocessing import StandardScaler

from sklearn.model_selection import train_test_split

from sklearn.linear_model import LinearRegression

from sklearn.linear_model import Ridge

from sklearn.linear_model import SGDRegressor

from sklearn.linear_model import Lasso

from sklearn.preprocessing import PolynomialFeatures

from sklearn.tree import DecisionTreeRegressor

from sklearn.ensemble import RandomForestRegressor

from sklearn.ensemble import GradientBoostingRegressor

from sklearn.neighbors import KNeighborsRegressor

from sklearn.svm import SVR

from sklearn.neural_network import MLPRegressor

from sklearn.preprocessing import PowerTransformer,
QuantileTransformer
```

```
from skopt import BayesSearchCV
from sklearn.metrics import mean_absolute_error, mean_squared_error
from sklearn.model_selection import cross_val_predict, cross_val_score, RandomizedSearchCV, Kfold


files = glob.glob(r"\*.parquet")
list = []
for filename in files:
    df = pd.read_parquet(filename)
    list.append(df)

data = pd.concat(list, axis=0, ignore_index=True)

data.head()
data.info()
data.isnull().sum()
```

Data preprocessing code

```
data['airport_fee'].fillna(data['airport_fee'].median(), inplace = True)
data['passenger_count'].fillna(data['passenger_count'].median(), inplace = True)
data['RatecodeID'].fillna(data['RatecodeID'].median(), inplace = True)
data['congestion_surcharge'].fillna(data['congestion_surcharge'].median(), inplace = True)

data['total_amount']=data['total_amount']-data['tip_amount']
data['trip_duration'] = (data['tpep_dropoff_datetime']-data['tpep_pickup_datetime'])/np.timedelta64(1, 'm')
```

```
data['pickup_hour'] = data['tpep_pickup_datetime'].dt.hour
data['day_of_week'] = data['tpep_pickup_datetime'].dt.weekday
data['month'] = data['tpep_pickup_datetime'].dt.month

data.drop({'VendorID','tpep_pickup_datetime','tpep_dropoff_datetime'
,'store_and_fwd_flag','payment_type','fare_amount','tip_amount','tolls_am
ount','Airport_fee'}, axis=1, inplace = True)

data.isnull().sum()
columns = data.columns.values
for col in columns:
    print(col,len(data[col].unique()))

data['passenger_count'].value_counts()
data['passenger_count'].values[data['passenger_count'] < 1] =
data['passenger_count'].median()

data['RatecodeID'].value_counts()
data['RatecodeID'].values[data['RatecodeID'] > 6] =
data['RatecodeID'].median()

data['congestion_surcharge'].value_counts()
data['congestion_surcharge'].values[data['congestion_surcharge'] <
0] = data['congestion_surcharge'].median()
data = data.loc[(data['congestion_surcharge']==0.00) |
(data['congestion_surcharge']==2.50)]

data['improvement_surcharge'].value_counts()
data['improvement_surcharge'].values[data['improvement_surcharge'] <
0] = data['improvement_surcharge'].median()
```

```
data['airport_fee'].value_counts()
data['airport_fee'].values[data['airport_fee'] < 0] = data['airport_fee'].median()

data['mta_tax'].value_counts()
data['mta_tax'].values[data['mta_tax'] < 0] = data['mta_tax'].median()
data = data.loc[(data['mta_tax']==0.5) | (data['mta_tax']==0)]

normalization = ['extra', 'trip_distance', 'trip_duration', 'total_amount']
for par in normalization:
    data[par].value_counts()
    data[par].values[par] < 0] = data[par].median()
    Q1 = data[par].quantile(0.25)
    Q3 = data[par].quantile(0.75)
    IQR = Q3 - Q1
    lower = Q1 - 1.5*IQR
    upper = Q3 + 1.5*IQR
    data = data.loc[(data[par]>=lower) & (data[par]<=upper)]
    data = data.loc[data[par]>0]

X = data.drop(columns=['total_amount'])
encoder = OneHotEncoder()
ratecode = encoder.fit_transform(data[['RatecodeID']]).toarray()
ratecode = pd.DataFrame(ratecode)
ratecode = ratecode.rename(columns = {0 : 1, 1 : 2, 2 : 3, 4 : 5, 5 : 6})
data = pd.concat([data, ratecode], axis=1)
data.drop({'RatecodeID'}, axis=1, inplace = True)
X = pd.get_dummies(X,columns=['RatecodeID'])
obj = StandardScaler()
X = obj.fit_transform(X)
y = data["total_amount"].values
X_train, X_test, y_train, y_test = train_test_split(X, y, test_size = 0.003)
```

Data analysis code

```
%matplotlib inline

sns.relplot(y='total_amount', x='trip_duration', data=data[::100], hue='RatecodeID', style='RatecodeID')


fig,ax1 = plt.subplots(figsize=(15,10))

ax = sns.countplot(x='pickup_hour', data=data, ax=ax1)


med_amount = data.groupby(['RatecodeID'])['total_amount'].median()

med_distance = data.groupby(['RatecodeID'])['trip_distance'].median()

med_duration = data.groupby(['RatecodeID'])['trip_duration'].median()


fig = plt.figure()

ax = fig.add_subplot(projection='3d')

fare = med_amount.values;

minutes = med_distance.values;

distance = med_duration.values;

lables = ["Standard rate","JFK", "Newark", "Nassau or Westchester","Negotiated fare", "Group ride"];

for i in range(6):

    ax.scatter(fare[i], minutes[i], distance[i])

ax.text(fare[i], minutes[i], distance[i], lables[i])


ax.set_xlabel('fare')

ax.set_ylabel('distance')

ax.set_zlabel('minutes')

plt.show()
```

Code for working with machine learning methods

```
model = LinearRegression().fit(X_train,y_train)
y_pred = model.predict(X_test)
print(mean_absolute_error(y_test, y_pred))
print(mean_squared_error(y_test, y_pred))
plt.hist(y_test, bins=np.arange(0, 30), alpha=0.3, color='blue', label='y')
plt.hist(y_pred, bins=np.arange(0, 30), alpha=0.3, color='red', label='y_prediction')
plt.legend(loc='upper right')
plt.xlabel('total_amount')
plt.show()

model = Ridge().fit(X_train, y_train)
y_pred = model.predict(X_test)
print(mean_absolute_error(y_test, y_pred))
print(mean_squared_error(y_test, y_pred))
plt.hist(y_test, bins=np.arange(0, 30), alpha=0.3, color='blue', label='y')
plt.hist(y_pred, bins=np.arange(0, 30), alpha=0.3, color='red', label='y_prediction')
plt.legend(loc='upper right')
plt.xlabel('total_amount')
```

```
plt.show()


model = Lasso().fit(X_train, y_train)
y_pred = model.predict(X_test)
print(mean_absolute_error(y_test, y_pred))
print(mean_squared_error(y_test, y_pred))
plt.hist(y_test, bins=np.arange(0, 30), alpha=0.3, color='blue', label='y')
plt.hist(y_pred, bins=np.arange(0, 30), alpha=0.3, color='red', label='y_prediction')
plt.legend(loc='upper right')
plt.xlabel('total_amount')
plt.show()


model = SGDRegressor(learning_rate='adaptive').fit(X_train, y_train)
y_pred = model.predict(X_test)
print(mean_absolute_error(y_test, y_pred))
print(mean_squared_error(y_test, y_pred))
plt.hist(y_test, bins=np.arange(0, 30), alpha=0.3, color='blue', label='y')
plt.hist(y_pred, bins=np.arange(0, 30), alpha=0.3, color='red', label='y_prediction')
plt.legend(loc='upper right')
plt.xlabel('total_amount')
plt.show()


model = DecisionTreeRegressor().fit(X_train, y_train)
y_pred = model.predict(X_test)
print(mean_absolute_error(y_test, y_pred))
print(mean_squared_error(y_test, y_pred))
```

```
plt.hist(y_test, bins=np.arange(0, 30), alpha=0.3, color='blue', label='y')
plt.hist(y_pred, bins=np.arange(0, 30), alpha=0.3, color='red', label='y_prediction')
plt.legend(loc='upper right')
plt.xlabel('total_amount')
plt.show()

model = RandomForestRegressor().fit(X_train, y_train)
y_pred = model.predict(X_test)
print(mean_absolute_error(y_test, y_pred))
print(mean_squared_error(y_test, y_pred))
plt.hist(y_test, bins=np.arange(0, 30), alpha=0.3, color='blue', label='y')
plt.hist(y_pred, bins=np.arange(0, 30), alpha=0.3, color='red', label='y_prediction')
plt.legend(loc='upper right')
plt.xlabel('total_amount')
plt.show()

params= {
        'criterion': ['squared_error', 'friedman_mse', 'absolute_error', 'poisson'],
        'n_estimators': [25, 50, 100, 150],
        'max_features': ['sqrt', 'log2', None],
        'max_depth': [3, 6, 9, 12],
        'max_leaf_nodes': [3, 6, 9, 12]
        }
model = RandomForestRegressor()
bayes = BayesSearchCV(model, params)
bayes.fit(X_train, y_train)
```

```
print("Best Hyperparameters:", bayes.best_params_)


model = RandomForestRegressor(n_estimators=350, criterion='friedman_mse', max_depth=None, min_samples_split=19, max_features=10, oob_score=True, verbose=350).fit(X_train, y_train)

y_pred = model.predict(X_test)

print(mean_absolute_error(y_test, y_pred))

print(mean_squared_error(y_test, y_pred))

plt.hist(y_test, bins=np.arange(0, 30), alpha=0.3, color='blue', label='y')

plt.hist(y_pred, bins=np.arange(0, 30), alpha=0.3, color='red', label='y_prediction')

plt.legend(loc='upper right')

plt.xlabel('total_amount')

plt.show()


model = GradientBoostingRegressor().fit(X_train, y_train)

y_pred = model.predict(X_test)

print(mean_absolute_error(y_test, y_pred))

print(mean_squared_error(y_test, y_pred))

plt.hist(y_test, bins=np.arange(0, 30), alpha=0.3, color='blue', label='y')

plt.hist(y_pred, bins=np.arange(0, 30), alpha=0.3, color='red', label='y_prediction')

plt.legend(loc='upper right')

plt.xlabel('total_amount')

plt.show()


params= {

        ' loss ': ['squared_error', ' huber', 'absolute_error', ' quantile'],
```

```
          'n_estimators': [200, 300, 400, 500],
          ' learning_rate ': list(np.arange(0.1, 0.9, 0.2)),
          ' min_samples_split': [2, 4, 6, 8, 10],
          ' min_samples_leaf':  [2, 4, 6, 8, 10]
          }
    model = GradientBoostingRegressor()
    bayes = BayesSearchCV(model, params)
    bayes.fit(X_train, y_train)
    print("Best Hyperparameters:", bayes.best_params_)

    model = GradientBoostingRegressor(loss='huber', learning_rate=0.7,
n_estimators=500, min_samples_split=10, min_samples_leaf=10,
verbose=500).fit(X_train, y_train)
    y_pred = model.predict(X_test)
    print(mean_absolute_error(y_test, y_pred))
    print(mean_squared_error(y_test, y_pred))
    plt.hist(y_test, bins=np.arange(0, 30), alpha=0.3, color='blue',
label='y')
    plt.hist(y_pred, bins=np.arange(0, 30), alpha=0.3, color='red',
label='y_prediction')
    plt.legend(loc='upper right')
    plt.xlabel('total_amount')
    plt.show()

    model = KNeighborsRegressor().fit(X_train, y_train)
    y_pred = model.predict(X_test)
    print(mean_absolute_error(y_test, y_pred))
    print(mean_squared_error(y_test, y_pred))
    plt.hist(y_test, bins=np.arange(0, 30), alpha=0.3, color='blue',
label='y')
```

```
    plt.hist(y_pred,   bins=np.arange(0,   30),   alpha=0.3,   color='red',
label='y_prediction')
    plt.legend(loc='upper right')
    plt.xlabel('total_amount')
    plt.show()

    model = SVR().fit(X_train, y_train)
    y_pred = model.predict(X_test)
    print(mean_absolute_error(y_test, y_pred))
    print(mean_squared_error(y_test, y_pred))
    plt.hist(y_test,   bins=np.arange(0,   30),   alpha=0.3,   color='blue',
label='y')
    plt.hist(y_pred,   bins=np.arange(0,   30),   alpha=0.3,   color='red',
label='y_prediction')
    plt.legend(loc='upper right')
    plt.xlabel('total_amount')
    plt.show()

    params= {
            '   kernel':   ['linear',   'poly',   'rbf',   'sigmoid',
'precomputed'],
            'C': list(np.arange(1, 21, 5)),
            'gamma': ['auto', 'scale'],
            'epsilon':  [0.1, 0.01, 0.001, 0.0001]
            }
    model = SVR()
    bayes = BayesSearchCV(model, params)
    bayes.fit(X_train, y_train)
    print("Best Hyperparameters:", bayes.best_params_)
```

```
model = SVR(C=10).fit(X_train, y_train)
y_pred = model.predict(X_test)
print(mean_absolute_error(y_test, y_pred))
print(mean_squared_error(y_test, y_pred))
plt.hist(y_test, bins=np.arange(0, 30), alpha=0.3, color='blue', label='y')
plt.hist(y_pred, bins=np.arange(0, 30), alpha=0.3, color='red', label='y_prediction')
plt.legend(loc='upper right')
plt.xlabel('total_amount')
plt.show()

model = MLPRegressor().fit(X_train, y_train)
y_pred = model.predict(X_test)
print(mean_absolute_error(y_test, y_pred))
print(mean_squared_error(y_test, y_pred))
plt.hist(y_test, bins=np.arange(0, 30), alpha=0.3, color='blue', label='y')
plt.hist(y_pred, bins=np.arange(0, 30), alpha=0.3, color='red', label='y_prediction')
plt.legend(loc='upper right')
plt.xlabel('total_amount')
plt.show()

params= {
        'solver': ['lbfgs', 'sgd', 'adam'],
        'hidden_layer_sizes': [(100,),(200,),(300,)],
        'activation': ['identity', 'logistic', 'tanh', 'relu'],
        'alpha':  [1e-7, 1e-6, 1e-5, 1e-4, 1e-3, 1e-2]
        }
```

```
model = MLPRegressor()
bayes = BayesSearchCV(model, params)
bayes.fit(X_train, y_train)
print("Best Hyperparameters:", bayes.best_params_)

model = MLPRegressor(hidden_layer_sizes=(200,100,50), activation='logistic', solver='adam', max_iter=500, verbose=True).fit(X_train, y_train)
y_pred = model.predict(X_test)
print(mean_absolute_error(y_test, y_pred))
print(mean_squared_error(y_test, y_pred))
plt.hist(y_test, bins=np.arange(0, 30), alpha=0.3, color='blue', label='y')
plt.hist(y_pred, bins=np.arange(0, 30), alpha=0.3, color='red', label='y_prediction')
plt.legend(loc='upper right')
plt.xlabel('total_amount')
plt.show()
```

Printed by Books on Demand GmbH, Norderstedt / Germany